Deivielison X. S. Macedo
Viviane C. Santos
Leonardo A. Monteiro

Initial soil preparation - Principles

Deivielison X. S. Macedo
Viviane C. Santos
Leonardo A. Monteiro

Initial soil preparation - Principles

Basic concepts

ScienciaScripts

Imprint
Any brand names and product names mentioned in this book are subject to trademark, brand or patent protection and are trademarks or registered trademarks of their respective holders. The use of brand names, product names, common names, trade names, product descriptions etc. even without a particular marking in this work is in no way to be construed to mean that such names may be regarded as unrestricted in respect of trademark and brand protection legislation and could thus be used by anyone.

Cover image: www.ingimage.com

This book is a translation from the original published under ISBN 978-3-330-75852-0.

Publisher:
Sciencia Scripts
is a trademark of
Dodo Books Indian Ocean Ltd. and OmniScriptum S.R.L publishing group

120 High Road, East Finchley, London, N2 9ED, United Kingdom
Str. Armeneasca 28/1, office 1, Chisinau MD-2012, Republic of Moldova, Europe
Printed at: see last page
ISBN: 978-620-8-30256-6

SUMMARY

1. INTRODUCTION TO INITIAL SOIL PREPARATION 3
2. PLANNING AND DEFORESTATION NO 12
3. MECHANIZED DEFORESTATION WITH FRONT BLADES 19
4. CURRENT DEFORESTATION 29
5. POST-DEFORESTATION ACTIVITIES 35
REFERENCES 41

Initial soil preparation: principles

AUTHORS:

Deivielison Ximenes Siqueira Macedo

Master in Agricultural Engineering - Federal University of Cearà

Viviane Castro dos Santos

Master in Agricultural Engineering - Federal University of Cearà

Leonardo de Almeida Monteiro

Professor - Federal University of Cearà

Aline Castro Praciano

Master in Agricultural Engineering - Federal University of Cearà

Eduardo Santos Cavalcante

Master in Agricultural Engineering - Federal University of Cearà

Enio Costa

Master Teacher - Federal Institute of Cearà

Jean Lucas Pereira Oliveira

Agronomy student - Federal University of Cearà

Jefferson Auteliano Carvalho Dutra

Agricultural Engineer

José Siqueira de Macedo Jùnior

Veterinary student - Northeast College of Technology

Karla Lùcia Batista Araùjo

Professor - Northeast Technology College

Maiara Pereira Barros

Agronomy student - Federal University of Cearà

1. INTRODUCTION TO INITIAL SOIL PREPARATION

Deivielison Ximenes Siqueira Macedo; Viviane Castro dos Santos; José Siqueira de Macedo Jùnior; Enio Costa; Maiara Pereira Barros

The world's population currently stands at around 7 billion people, and experts are faced with the challenge of how to feed them. The main ways of achieving this goal are by increasing the productivity of existing areas, restoring degraded areas, restoring abandoned areas and opening up new areas.

Brazil has around 152.5 million hectares of arable land, corresponding to approximately 17.5% of the national territory, but only 62.5 million hectares are used, meaning that there are still 90 million hectares unused, approximately 10.5% of the national territory, i.e. new areas to be cleared. So, in order to take advantage of these areas that have not yet been exploited, they are being initially prepared.

Initial soil preparation can be defined as: the set of operations necessary to create the conditions for planting crops in areas not previously used for this purpose.

Figure 1: Initial preparation of an area.

Source: O CORRENTÂO (2016)

1.1 Initial preparation stages

The initial preparation is basically divided into 4 phases:

- Deforestation
- Clearance
- Thickening
- Leveling

Deforestation

There is a lot of talk about the predatory action of deforestation and its harmful effects on nature and the soil, but is it really deforestation? Is deforestation the same thing as devastation? Many people ask themselves these questions and don't know the answer for sure.

Deforestation is a legal process that must provide for the planning for which the deforested area will be used. The operation itself must create arable land, the coverage of which with artificial vegetation must in the short term replace the natural one, i.e. "bare" land is not allowed.

Figure 2: Legal deforestation, conservation units and permanent preservation areas in the São Francisco Basin.

Source: CI FLORESTAS (2016).

In contrast, deforestation is the removal of vegetation without any prior study, with the aim of making a total profit from exploiting the area to its detriment. In other words, it is the removal of vegetation with the aim of exploiting timber, without concern for the soil or what will happen to the area after it has been removed.

Figure 3: Devastated area.

Source: CEARÀ AGORA (2016).

In this way, there is a clear difference between the concepts: one is studied, prepared with purpose and the other is loose, one is legal and the other illegal. Some of the tactics of each operation give a good account of how and when these concepts are practiced.

According to Alencar *et al.* (2004) the practice is considered (Legal) Deforestation when:

- It complies with current legislation and is carried out in soil and climate that is suitable for farming and therefore productive;

- It occurs in areas with adequate infrastructure and access to markets, and therefore with a low risk of early abandonment of the economic activity being established;

- It is carried out in areas with a high density of established rural populations who can benefit from forest conversion, for example by increasing social equity;

- It occurs where traditional/indigenous populations are dependent on subsistence farming.

As for (Illegal) Devastation, according to Alencar *et al.* (2004) when:

- It only aims to justify land ownership and is used speculatively to "show productivity" to government agencies such as INCRA;

- It occurs on land unsuitable for growing crops and raising livestock (e.g. rugged terrain and unsuitable soils), leading to low productivity farming systems;

- When it does not comply with the Forest Code;

- It occurs in conservation units, indigenous lands, or in areas of high value for the conservation or sustainable use of biodiversity;

- It occurs in areas where the best economic option for land use is forestry - either for timber or non-timber production, or both.

One of the premises for carrying out deforestation is to comply with the rules in force in the country, such as the Forestry Code. According to the code, engineers must pay attention to the Areas of Permanent Preservation (APPs) and the Legal Reserve Area (ARL) when opening new areas. The definitions and information on PPAs and LRAs are in accordance with the Forestry Code. This book covers only a small part of what the law says. If you need more information, you should consult the forestry code as a whole.

1.2 Permanent Preservation Area - APP

According to Article 3 of the Law, a **permanent preservation** area is a protected area, whether or not it is covered by native vegetation, with the environmental function of preserving water resources, the landscape, geological stability and biodiversity, facilitating the gene flow of fauna and flora, protecting the soil and ensuring the well-being of human populations;

Figure 4: Permanent Preservation Area - APP

Source: ALY SOUSA (2016).

According to Article 4 of the Forest Code, the following are considered **Permanent Preservation Areas**, in **rural or urban areas**: I - the marginal strips of any natural perennial or intermittent watercourse, excluding ephemeral watercourses (An ephemeral body of water, stream, river, lake or pond is one that only exists for a few days after a rainfall or thaw). They are not the same as intermittent or seasonal bodies of water that exist for longer periods, but not all year round, from the edge of the regular bed channel, at a minimum width of:

- At the source, 50 meters of forest should be left around the river;
- If the body of water is 10 meters wide, leave 30 meters of forest around the body of water;
- If the body of water is 10 to 50 meters wide, leave 50 meters of forest around the body of water;
- If the body of water is 50 to 200 meters wide, 100 meters of forest is left around the body of water;
- If the body of water is 200 to 500 meters wide, leave 200 meters of forest around

the body of water;

- If the body of water is more than 600 meters wide, leave 500 meters of forest around the body of water.

Figure 5: Minimum width of APP according to the width of water bodies.

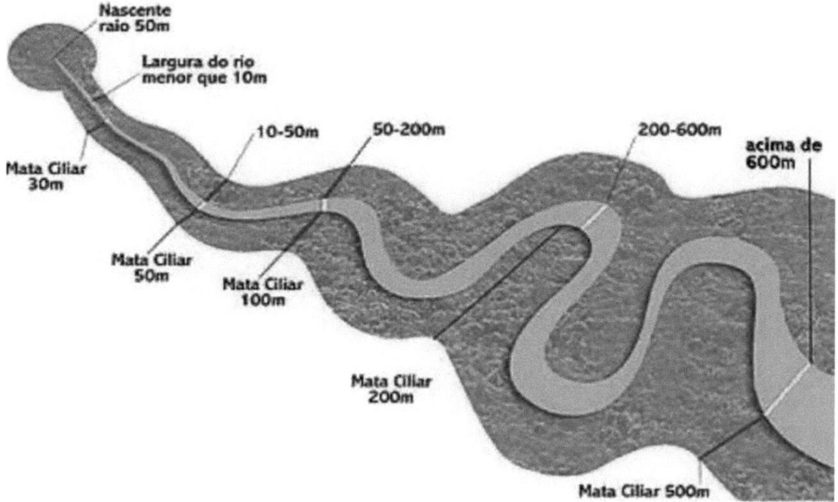

Source: WWF (2016)

Riparian forest is the vegetation on the banks of bodies of water such as rivers, streams and lakes. The maintenance of this vegetation around these bodies is intended to protect the areas of permanent preservation from the source to the mouth.

According to Article 4 of the law, item II: the areas around artificial water reservoirs, in the range defined in the environmental license for the project, subject to the provisions of paragraphs 1, 2 and 4.

§ Paragraph 1. The provisions of item III shall not apply in cases where artificial water reservoirs are not the result of damming or impounding watercourses.

§ Paragraph 2 - In the surroundings of artificial reservoirs located in rural areas of up to 20 (twenty) hectares, the area of permanent preservation will be at least 15 (fifteen) meters.

§ Paragraph 4 - In natural or artificial accumulations of water with a surface area of less than 1 (one) hectare, the reservation of the protection strip provided for in items II and III of the caput is dispensed with.

Figure 6: APP in a real situation.

Source: DANIELA KREIN (2013).

1.3 Legal reserve

According to Article 3 of the Law, a legal reserve is defined as:

III - Legal Reserve: an area located within a rural property or possession, delimited under the terms of art. 12, with the function of ensuring the sustainable economic use of the rural property's natural resources, assisting in the conservation and rehabilitation of ecological processes and promoting the conservation of biodiversity, as well as the shelter and protection of wild fauna and native flora.

Figure 7: Legal Reserve areas.

Source: MUNDO AMBIENTAL (2010).

Article 12 of the Law states that the legal reserve is obligatory - Every rural property must maintain an area covered with native vegetation, without prejudice to the application of the rules on permanent preservation areas, observing the following minimum percentages in relation to the area of the property:

I - located in the legal Amazon (Amazon free for exploration):

a) 80% on property located in a forest area;

b) 35% on property located in a savannah area;

c) 20% on a property located in an area of general fields.

- II - located in other regions of the country: 20%.

Article 3 of the Law treats the Legal Amazon as: I- Legal Amazon: the states of Acre, Parà, Amazonas, Roraima, Rondônia, Amapà and Mato Grosso and the regions situated to the north of the 13° S parallel, in the states of Tocantins and Goiàs, and to the west of the 44° W meridian, in the state of Maranhao.

Figure 8: Legal Reserve in Forests.

Source: MAURO GUANADI (2009).

EXERCISES

1. What are the ways to increase food production in the world?

2. What is initial soil preparation and what are its stages?

3. What is the difference between deforestation and devastation?

4. What is a Permanent Preservation Area and a Legal Reserve? Are there any differences between them?

5. What is riparian forest?

6. What size riparian forest should be left in an APP?

7. What factors will influence the size of the Legal Reserve area?

2. PLANNING AND DEFORESTATION NO MECHANIZED

Deivielison Ximenes Siqueira Macedo; Viviane Castro dos Santos; José Siqueira de Macedo Jùnior; Jefferson Auteliano Carvalho Dutra; Maiara Pereira Barros

Deforestation, as discussed above, is a legal technique that aims to remove native or artificial vegetation from an area in order to plant a crop of interest. To this end, certain factors must be taken into account before it begins.

2.1 Factors to take into account in deforestation

Deforestation can be considered an art when it is carried out correctly and precisely, since it can only be done with a prior study of the area, where the actions taken will depend on a number of factors present in the region. The factors that most influence decision-making are vegetation, soil, climatic conditions, the purpose of the land use and the weather.

a) Vegetation

The shape of the vegetation should be checked, whether it is specific or forest.

In the specific form, the trunk is robust, the crown is proportional to the trunk and the tree is not so large, as there is no competition for light with other plants. *In the* forest form, the trunk is thinner, the crown is smaller and the height of the plant is greater, as there is competition between the plants for light.

This is one of the most important factors to take into account during initial preparation, as it is through the vegetation that the choice of machinery is made.

b) Soil

You should check the type of soil (sandy or clay), whether the soil profile is deep or there are rock outcrops in the area, the topography of the land in relation to the slope (in the case of very steep slopes, it may be impossible to use machinery to carry out the work).

c) Climatic conditions

The period of the year must be taken into account, because in months with a surplus of rain it can affect the operation, flood the soil and even damage the machines through prolonged exposure to unsuitable climatic conditions, reducing their useful life.

d) Purpose of land use

The use of the land also influences the initial soil preparation operations. If the land is to be used for agriculture, all vegetation should be removed from the area where the crop is to be planted, including stumps. In the case of roots, they should be removed at a depth slightly below the depth of the root system of the crop of interest, i.e. if the crop to be planted after initial preparation has a root system of around 30 cm, it is not necessary to remove the roots up to a depth of 1 meter.

If the land is destined for livestock farming, the engineer doesn't need to remove all the trees, but can leave a few in strategic places to shade the cattle.

e) Time

The time factor is directly related to the investment factor: the faster you want the area ready for planting the crop of economic interest, the greater the investment in the equipment to be used for preparation.

2.2 Design and Planning

The design and planning part is where the area is evaluated and how and when it will be prepared is defined, taking into account mainly the factors mentioned above.

There are a number of ways of surveying the vegetation in the area, the best known being the square method, the strip method and the use of maps, photos and satellites.

a) Square

The engineer must carry out 3 or 4 counts in small areas of 0.1 ha and estimate for the rest of the area. The choice is made at random and this method is the least precise compared to the others.

b) Track

This method is more reliable than the quadrat method because it assesses all the vegetation in the area, i.e. for each different type of vegetation found in the area to be cleared, 3 or 4 samples are taken. The survey of the area is also random and 0.2 ha is estimated for each count.

c) Maps, photos and satellites

Some satellites can be used to take aerial photographs of certain regions, which are then processed in software to make maps available. Another way is to take aerial photographs using drones.

Figure 9: Images obtained from aerial photography and map estimating the vegetation of the site.

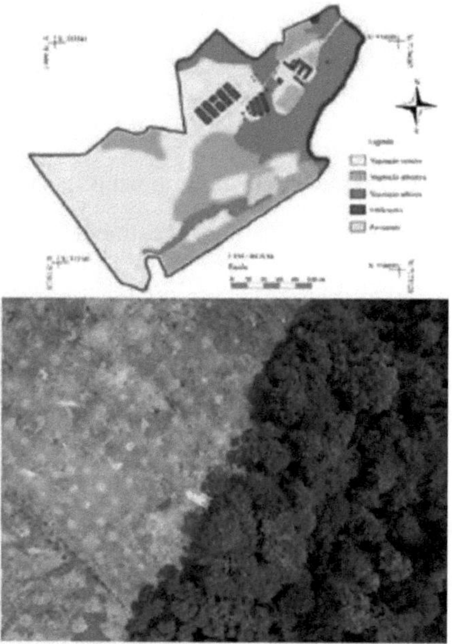

Source: VENANCIO *et al* 2010 and MONGABAY (2014).

Data collected in the project and planning

When you count the trees, you should count all the trees present in each sample:

1. Density of vegetation less than 30 cm in diameter:

Dense - more than 1500 trees/ha

Average - from 1000 to 1500 trees/ha

Rala - less than 1000 trees/ha

2. Presence of hardwood expressed as a percentage;

3. Presence of vines;

4. Sum of all trees/ha over 1.80 m in diameter at ground level;

5. Average number of trees/ha, according to diameter at ground level, classified as follows:

- Less than 30 cm;
- From 30 cm to 60 cm;
- From 60 cm to 90 cm;
- From 90 cm to 120 cm;
- From 120 to 180 cm.

2.3 Non-mechanized deforestation

Non-mechanized deforestation can be done by hand, animal or chemical means.

Manual deforestation is done using chainsaws, axes, hoes and other materials that help with cutting or digging. The engineer must take into account the size of the area, the diameter of the trunk and the root system of the plant. Large areas and plants with very large diameters are not viable for manual felling. For trees with a fasciculated root system, you should first dig up the roots, cut them off and then cut down the plant. For trees with a pivoting root system, you should cut the trunk closest to the ground and proceed with the use of desiccants (potassium nitrate or quicklime) to wilt the trunk and roots.

Figure 10: Difference between root systems.

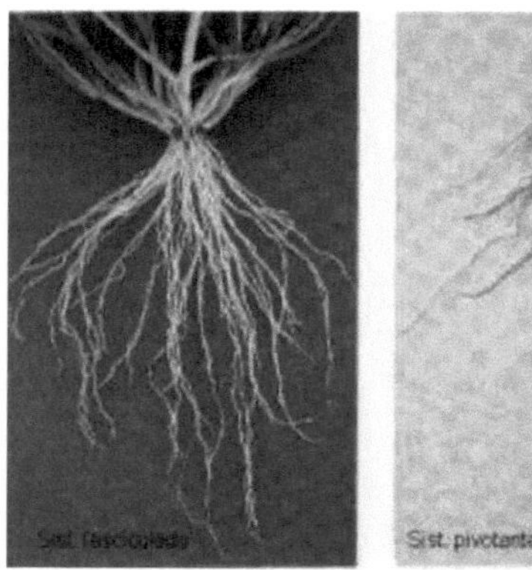

Source: SOBIOLOGIA (2016).

The best-known form of animal logging is the whip method, in which chains are tied between trees and the animals are led around until they twist the tree.

Figure 11: Mule, one of the animals used for the method.

Source: COMPRE RURAL (2016).

Figure 12: Chain around the tree to be felled.

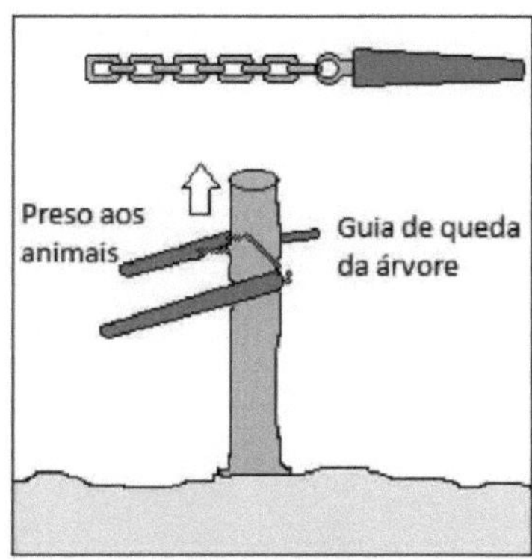

Source: MACEDO (2016).

Deforestation with chemicals is one of the most time-consuming and is greatly influenced by the time of year. It can take up to 6 months to fell a tree. Generally, chemical products are used on isolated trees, where it is feasible to use them. To carry it out, open a slit in the trunk near the conducting vessels and deposit ammonium nitrate or other nitrogenous substances, replenishing as necessary; after a while the tree will wither and rot and can be easily removed.

EXERCISES

1 What factors are taken into account before deforestation begins?

2 How can each factor influence deforestation?

3 What methods are taken into account when planning and designing a deforestation project?

4 What is the difference between the strip method and the square method?

5 What data should be collected when exploring an area?

6 What are the types of non-mechanized deforestation?

7 What factors should be taken into account when clearing land manually?

8 How should trees with pivoting root systems be removed?

9 How is animal deforestation carried out?

10 When and how is chemical deforestation carried out?

3. MECHANIZED DEFORESTATION WITH FRONT BLADES

Deivielison Ximenes Siqueira Macedo; Viviane Castro dos Santos; Aline Castro Praciano; Karla Lùcia Batista Araùjo; Enio Costa; Jean Lucas Pereira Oliveira

Mechanized deforestation can be carried out using two types of equipment: the first is front blades and the other is chains. Front blades are used in smaller areas, but in areas with vegetation of a more robust diameter and with a high density of vegetation, such as forests or isolated vegetation, while the use of chains is recommended for larger areas with vegetation of a thinner diameter, and is ideal for savannahs and woodlands.

The blades are generally attached to the front of wheeled or tracked tractors and are classified as smooth or pushing and cutting or angled.

3.1 Smooth front blades

There are two smooth front blades for deforestation: the Angledozer and the Bulldozer. These blades have no spurs or sharp cutting edges, are mounted on the front of tracked tractors or wheeled tractors and are used for light, medium or heavy deforestation and were initially designed for cutting, transporting and leveling.

Figure 13: Smooth or pushing front blades.

Source: CLASSIC DOZERS (2016).

Bulldozer

Bulldozer blades, compared to angledozer blades, are said to be fixed and are used for earthmoving. They are longer and hollower than angledozer blades and have side feet that prevent the loaded material from flowing sideways.

Figure 14: Tractor with bulldozer blade

Source: XLHJ (2016).

It is more robust and is mainly used for heavy-duty work, in harsh conditions, and carries a larger volume. It moves up and down and has a hydraulic system which makes small movements in its forward angle.

Figure 15: Moving the bulldozer blade.

Source: COSTO JUSTO (2016).

Angledozer

Compared to the bulldozer, the Angledozer Smooth Blade is smaller, with less concavity, no side feet, but wider. It is not very suitable for transporting materials over long distances, more than 50 meters, because due to the absence of side feet and the low concavity, the loaded material tends to flow sideways, making it uneconomical.

Figure 16: Angledozer blade

Source: JIMAYA (2009).

Its movement is different from the bulldozer, being forward and backward and up and down at a single point.

Figure 17: Moving the angledozer for earthmoving.

Source: VITOR CARNOTO (2016); NEW ROLAND (2009).

Due to these angles, the angledozer can be used for surface cleaning of easily removable rubble, planning and transporting earth, as well as for cutting embankments, as well as for road construction and building 50 cm deep drainage rods in marshy soils.

Cutting procedure for flat blades

This procedure is used on both flat blades, the main roots must be dug up and then cut, after which a force must be applied against the trunk to make it fall. It is important to note that before the tree falls, the operator must move away with the tractor to avoid being hit. After the tree falls, the earth is moved.

Operating capacity of smooth blades

The time these blades take to clear 1 ha varies between 20 and 30 hours. What will influence this is the power of the tractor, the density of the vegetation, the type of vegetation and the diameter of the trees, as shown in the table.

Table 1: Operating capacity of smooth blades as a function of tractor power and vegetation density.

CLASSES	Density of 500 bushes per m^2	Density of 1000 bushes per m2	Density of 1500 bushes per m2
75 to 100 HP	1 ha/25h, shrubs and trees with a diameter of 20 cm	-	-
100 to 150 HP	1 ha/20h with 5 trees 50 cm in diameter	1 ha/25h with 7 trees up to 50 cm in diameter	1 ha/30h with 10 trees over 50 cm in diameter
150 to 200 HP	1 ha/20h with 8 trees 50-60 cm in diameter	1 ha/25h with 10 trees 50-60 cm in diameter	1 ha/30h with 10 trees up to 1m in diameter
more than 200 HP	-	1 ha/20h with 10 trees 50-60 cm in diameter	1 ha/30h with 10 trees up to 1.5 m in diameter

Source: TESTA (1983).

3.2 Front Cutting Blades

The best-known front cutting blades are two: the Kg and the V-blade. They are specially designed for clearing vegetation, unlike smooth blades which were developed for mowing and moving earth. Their operational performance is 40% higher than that of smooth blades (SAAD, 1977).

Front felling blades are used in forests for the rapid removal of trees that do not sprout, generally leaving smooth areas with stumps close to the ground, but they are not recommended for rocky terrain.

Figure 18: Cutting blade.

Source: ROME (2016).

The two cutting blades include a spur or knife, a cutting bar and a deflector or guide bar.

S **Spur or knife** - to split and destroy the resistance of the trunk;

S **Cutting bar** - comes into operation after the stem has been removed;

S **Deflector or guide bar** - guides the fall of the spindle forward and to the right of the operator.

Figure 19: Cutting blade construction.

Source: KEASLER (2008) adapted by MACEDO (2016).

Front cutting blade Kg

This blade is the most common among the cutting blades. It is used for felling trees without the need to dig up the roots. It is also used for girdling, a post-deforestation process, due to its large hollow surface, but it does not remove materials below the ground (stones). Due to its cutting elements, it is avoided in order to remain sharp.

Figure 20: Blade kg.

Source: ROME (2016).

V-blade

This blade is arranged so that its spur is in the center, between the two plates that make it up, making it easy to remove. It is only used for quick felling, without being used for clearing and pollarding, as it has no concavity and has no way of pushing the cut material, i.e. pollarding the plant remains.

These blades are slightly wider than kg blades and are ideal for artificial forests, where the spacing between plants is already pre-defined, but these blades require more powerful tractors.

Figure 21: V-blade

Source: ROME (2016)

Auxiliary blades

Often in isolated areas there are tall trees with sturdy trunks, which would require a ramp to fell them so that the tractor would be high enough to reach the tree at a point where less force is needed to fell it. However, these ramps become unfeasible due to the intense revolving of the soil and the loss of the surface horizons (the best chemically for the plant), with this in mind, the fleco arm was developed.

Figure 22: Fleco arm.

Source: HOOPERS ENGINEERING (2016).

The fleco arm, also known as a pusher boom or tree boom, consists of a boom used to push and fell single trees. It is attached to the front of the crawler tractor and can reach a height of more than 4.57 m, forming a lever effect on the strongest trees.

It is generally used on very strong, tall trunks or trunks with a large number of roots that cannot be reached by the front blades or cannot be knocked down by the chains.

Figure 23: Pusher boom attached to the tractor along with a bulldozer.

Source: HOMAN (2016).

EXERCISES

1. What types of equipment are used for mechanized deforestation and how do they differ?

2. What is the difference between smooth blades and cutting blades?

3. What are the types of smooth front blades and how do they differ?

4. What are the procedures for cutting down trees with flat blades?

5. How long does it take to clear 1 ha using a smooth front blade?

6. What are the types of front cutting blades and how do they differ?

7. Why aren't front-cutting blades recommended in rocky areas?

8. What are the components and function of each of the front cutting blades?

9. What is the function of the fleco arm?

10. Under what conditions is the fleco arm used?

4. CURRENT DEFORESTATION

Deivielison Ximenes siqueira Macedo; Viviane castro dos santos; Aline Castro Praciano; Eduardo Santos Cavalcante; Karla Lùcia Batista Araùjo.

The correntao is a piece of equipment used mainly on huge properties with hundreds of hectares or not very dense vegetation. This equipment is nothing more than chains equipped with extremely resistant links with special links to engage and disengage small pieces and other rotating links to prevent them from winding up.

Figure 24: Chains

Source: E-SIRKET (2016).

The chains are attached to the back of the tractors and 2 or even 3 tractors are used in the felling process. The distance between the tractors must not be greater than 1/3 of the length of the chain. During the operation, the tractors must run symmetrically to avoid one pulling the other or working harder than the other. The chains are adjusted using cutting and welding tools. The tractors used to pull the chain range from 140 to 350 hp. In terms of productivity, operations with the correntao are much faster; to prepare 1 hectare, smooth front blades take 20 to 30 hours, at a rate of 0.03 to 0.05 hectare per hour, while the correntao deforests 4 to 28 hectares per hour.

Figure 25: Chaining on the tractor's rear axle.

Source: IBAMA (2011).

Still on the subject of chainsaws, they are normally used in two passes, the first of which is designed to bring down the vegetation, and the second, in the opposite direction to the first, is designed to uproot the vegetation and thinner trunks.

Figure 26: Trolling activity.

Source: CLASSIFICADOS AGRÌCOLA (2016).

Chains must be 3 times longer than the distance between tractors and have swivel shackles installed at least every 30 meters to reduce twisting.

Figura 27: Swivel links

Source: MF RURAL (2016).

To get good results with correntao, it should be used in places where the vegetation is cerrado or cerradôes, the area should have no more than 2,500 plants per hectare and the land should be well-drained, flat or slightly sloping.

Figura 28: Schematic drawing of chain operation

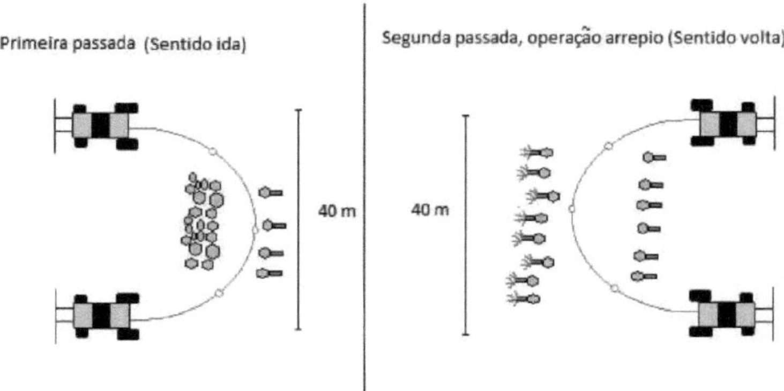

Source: MACEDO (2017).

Metal balls are used to improve the performance of chains or when felling vegetation requires a lot of tractor power. They are used to give the chain more weight and impact, forcing the chains to adapt to the shape of the terrain. Due to the height of the metal ball, they help to leverage and fell trees, but they should not be used in wet or swampy terrain.

Figure 29: Metal ball

Source: BANG SHIFT (2016).

More than one ball can be used per chain. If there is more than one ball, the distance between them must be equal, and if there is only one ball, it must be in the middle of the total length of the chain. The balls vary in diameter from 0.90 m to 1.80 m. They can be hollow or filled with concrete and can weigh between 2,500 and 7,500 kg.

Figure 30: Location of the metal float on the chain

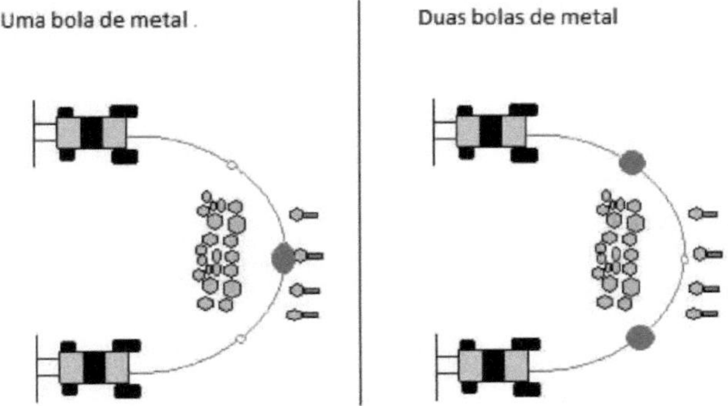

Source: MACEDO (2017).

The rake does not change the structure of the soil, unlike front blades, which remove soil when cutting or loading material. In the creep operation, which is the second pass, it leaves only small holes due to the stem coming out and in uprooting it leaves small cracks due to the removal of the roots.

Figure 31: Metal ball attached to the trolley.

Source: REVEGETATION EQUIPMENT CATALOG (2016).

The operation cannot be carried out on slopes greater than 5%, the vegetation must be homogeneous in terms of diameter, the height does not interfere with the operation, and shrubs and undergrowth do not pose any difficulties, being removed in the first pass.

EXERCISES

1. When do you use chains?
2. What is the composition of the chains?
3. Describe the correntao activity step by step.
4. Why is running more productive than smooth front blades?
5. What is a shivering operation?
6. What are the metal balls for?
7. How should the position of the metal balls be distributed on the chain?
8. What are the swivel shackles for and how are they installed on the chain?

5. POST-DEFORESTATION ACTIVITIES

Deivielison Ximenes Siqueira Macedo; Viviane Castro dos Santos; Jefferson Auteliano Carvalho Dutra; Eduardo Santos Cavalcante; Jean Lucas Pereira Oliveira

Post-deforestation activities basically consist of complementary activities to deforestation that make it possible to prepare the soil periodically afterwards. These are the other activities involved in the initial preparation of the soil, such as: removing stumps or clearing the area, removing spare stones (if any and if necessary), harrowing and leveling the soil.

5.1 Clearance

Clearing is the next stage after deforestation. Generally, after the area has been cleared and due to the use of some blades, trunks are left lying around. Before mechanized soil preparation, they and their roots must be removed, but at depths that are compatible with the work of the soil preparation machines.

Figure 30: Log left in the area after deforestation.

Source: TERRAMOV (2016).

The main pieces of equipment used in deforestation are the dethatcher, a tracked tractor with a smooth front blade (angledozer or bulldozer), a tracked tractor with a cutting blade (KG) and a dethatching blade.

Clearance consists of 3 distinct phases:

1. Preparing the stump - involves digging up and cutting off the main roots of the stump ("stripping" the stump);

2. Pulling - successive impacts against the stump, causing it to topple over and be completely removed;

3. Uprooting - removing the remaining roots with an uprooting rake or other tool.

After the natural vegetation has been felled and cleared, the material can still be used by cutting it into 0.90 to 1.0 m pieces using chainsaws. The wood from this type of vegetation is generally used for firewood or charcoal.

5.2 Thickening

This is the penultimate stage of the initial preparation and consists of piling up all the cut plant material away from the area, i.e. clearing the area. It takes place after the vegetation has been cut down and is done in such a way as to occupy the smallest possible area of the cut.

Figure 31: Thinning of vegetation after deforestation and clearing.

Source: THORCO (2017).

For economic reasons, to preserve machinery and fuel, several beds are laid out within the area, but some factors affect the spacing of these beds, such as the slope of the land, the amount of woody material to be gathered and the equipment used. Normally the beds are spaced 30 to 100 meters apart, depending on the area.

Girdling can be done in two ways and what will affect these ways is the equipment used to fell the trees. The first occurs at the time of felling, where it is done as the trees are being felled. This is possible with the use of KG cutting blades, which are suitable for both operations. The second is after felling, where it is applied when specific felling equipment is used, such as the correntao.

Figure 32: The wooded area in the cerrado.

Source: O CORRENTAO (2016).

The equipment most commonly used for threshing is:

- Smooth front blade (bulldozer or angledozer) - this is the least suitable piece of equipment, as it drags a lot of soil into the bed, scrapes the topsoil and can make ant-fighting more difficult.

- Front cutter blade (KG) - when properly adjusted, this blade gathers up the fallen material without scraping the topsoil;

- Front blade cultivator (cultivator rake) - this is the most suitable equipment, as it is specifically designed for this purpose.

Figure 33: Rolling rake.

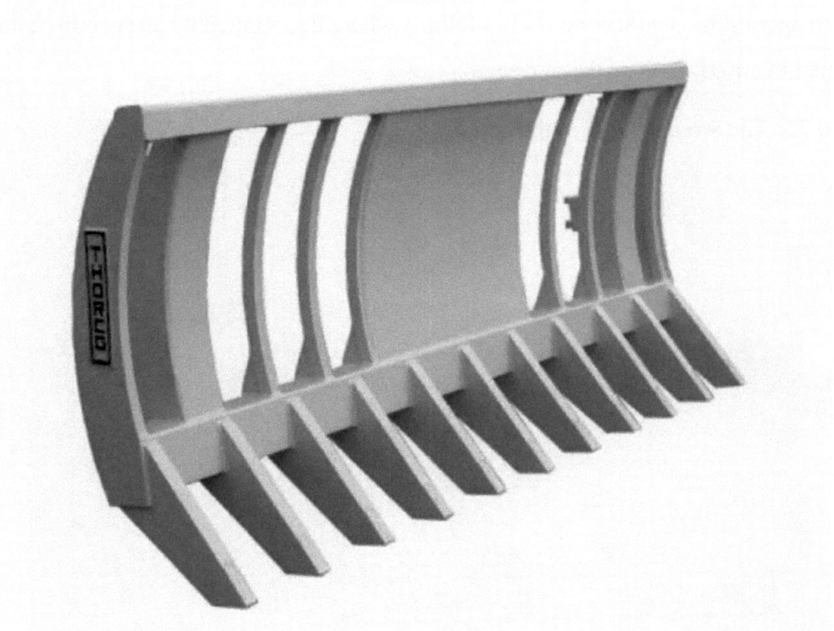

Source: THORCO (2017).

5.3 Leveling

Leveling the soil is the last phase of initial preparation. It is carried out when there are depressions, ditches, slopes, among other obstacles that make the surface uneven, hindering the establishment of the crop of interest or the periodic preparation to be installed in the area. Leveling the land, therefore, is an important practice that must precede soil preparation.

Leveling can be done manually using hoes, shovels, pickaxes and other tools, or mechanized, where the most commonly used equipment is the leveling planer and heavy harrows.

Figure 34: Leveling planer

Source: INPAL (2017).

5.4 Equipment according to vegetation and area.

At the end of this chapter, a summary can be made of the use of each piece of equipment according to the size of the area to be worked on and the diameter of the vegetation, as shown in Table 3.

Table 3: Summary of possible operations depending on vegetation and area size.

AREA	Vegetation 5 cm in diameter	Vegetation from 5 to 20 cm in diameter	Vegetation 20 cm or more in diameter
Small 4ha	Blades, Axes, Hoes and Chainsaws	Blades	Blade
Average 40ha	Blades	Blades	Cutting blade, Fleco arm, Rakes and
Large 400ha	Blades, Rakes and Chains dragged by 2	Angle cutting blade, Rakes and	Grinder

	crawler tractors	Correntao with 2 crawler tractors	Cutting blade, fleco arm, rakes, trimmer and steel ball chain pulled by 2 tractors

Source: TESTA (1983) adapted by MACEDO (2016).

EXERCISES

1. What are post-deforestation activities?
2. What is clearing?
3. What equipment do you use to cut down trees?
4. What are the stages of deforestation?
5. What does rooting consist of?
6. What are the forms of girdling and the factors that influence this activity?
7. What equipment is most commonly used for crimping?
8. What is leveling, when can it be done and what equipment is used to do it?

REFERENCES

ALENCAR, A.; NEPSTAD, D.; MCGRATH, D.; MOUTINHO, P.; PACHECO P., DIAZ, M. D. C. V.; SOARES FILHO, B. Deforestation in the Amazon: going beyond the "chronic emergency". IPAM, Parà, 89 p., 2004.

ALY SOUSA BLOG. Find out what APP. Online. Available at: < https://alysouza.files.wordpress.com/2011/02/kdk_0376.jpg> Accessed on Oct. 14, 2016.

BALLONI, E. A.; SIMÔES, J. W. Implementation of forest stands with species of the genus eucalyptus. Technical Circular nº 60, IPEF, 14 p., 1979.

BANG SHIFT. Best Bang shift: Bulldozers and steel balls - Bangshift Greatness circa 1950. Online. Available at: < http://bangshift.com/general- news/bulldozers-and-steel-balls-bangshift-greatness-circa-1950/> Accessed on: October 31, 2016.

BRAZIL. Forest Code. Law No. 12.651 of May 25, 2012.

CARNOTO, VITOR. Trade in agricultural and industrial machinery (used). Online. Available at: < http : //www.vitor-carnoto .com/products/bulldozer- caterpillar-lgp-d3b1/> Accessed on Oct. 16, 2016.

CEARAAGORA . OnlineAvailable : http://www.cearaagora.com.br/site/2015/08/brasil-reduz-em-15-o- deforestation-in-amazonia-legal/. Accessed on October 14, 2016.

CIFLORESTAS . Online. Available :< http://www.ciflorestas.com.br/arquivos/c_44_44_1306.jpg>. Accessed on October 14, 2016.

CLASSIC DOZERS. Information about classic crawlers and dozers. Online. Available at: < https://classicdozers.wordpress.com/dozer-gallerv/caterpillar- 1950-1970/>. Accessed on: October 16, 2016.

AGRICULTURAL AGRICULTURAL. Online. http://www.classificadosagricolas.com.br/produtos/maquinas/implementos/corre

ntao-agricola-rolo-corrente. Accessed on October 28, 2016.

BUY RURAL. Find out all about donkeys and mules and why professionals have been using this type of animal at work. 2016. Online. Available at: < http://www.comprerural.com/conheca-tudo-sobre-burros-e-mulas-e-porque-professionals-have-used-this-type-of-animal-at-work/>. Accessed on: October 15, 2016.

FAIR COST. Bulldozer caterpillar. Online. Available at: < http://www.custojusto.pt/guarda/outrosveiculos/bulldozer-caterpillar- 21008782>. Accessed on: October 16, 2016.

ENVOLVERDE. Online. Available at: <http : //envolverde.com.br/ambiente/desmatamento-ambiente/soj a-valorizada-e-expectativa-de-anistia-incentivam-desmatamento/> Accessed on July 7, 2014.

E-SIRKET. Online. Available at: < http://www.e-sirket.com/gemi-zinciri-16229e3.html>. Accessed on October 28, 2016.

GUANADI, MAURO. 2009. Online. Available at: < https://www.flickr.com/photos/mauroguanandi/4201117445/in/photostream/>. Accessed on: October 15, 2016.

HOMAN. Canopies and Treespears. Online. Available at: < http://www.homan.com.au/images/nseriescat%20full.jpg>. Accessed on October 28, 2016.

HOOPERS ENGINNERING. Catalogue tree spears. Online. Available at: http://www.hoopersengineering.com/catalogue/tree-spears. Accessed on October 28, 2016.

BRAZILIAN INSTITUTE FOR THE ENVIRONMENT AND NON-RENEWABLE NATURAL RESOURCES-IBAMA. 2011. Online. Available at: http://www.ibama.gov.br/publicadas/ibama-apreende-cinco-tratores-em- desmate-com-correntao-no-para. Accessed on October 28, 2016.

IDO, O. T.; OLIVEIRA, R. A. Land adaptation. Paranà, UFPR, 10 p. Online. Available at: < http : //www.agriculturageral. ufpr. br/biblio grafia/apostila2.pdf >. Accessed on Aug. 21, 2012.

BRAZILIAN INSTITUTE OF GEOGRAPHY AND STATISTICS - IBGE. Online. Available at: < http://www.ibge.gov.br/home/>. Accessed on July 7, 2014.

INPAL. Online. Available :< http : //www.inpal-to .com.br/portal/wp-content/uploads/2012/11/111.jpg>. Accessed on: 31 Jan. 2017.

JIMAYA .2006 .Online. Available : http://maquinariapesadaensinaloa.blogspot.com.br/2009/11/d6.html. Accessed on October 28, 2016.

KEASLE .2008 .Online. Available :< http://www.keaslersjunk.com/images 2008/rome kg dozer blade001 .jpg<. Accessed on: October 23, 2016.

KREIN, DANIELA. The importance of conservation areas. 2013. Online. Available at: < http : //ecolo ggando. blo gspot.com.br/2013/03/a-importancia- das-unidades-de.html>. Accessed on: October 15, 2016.

MFRURAL. Online. Available :< http://www.mfrural .com.br/detalhes.asp?cdp=179226&nmoca=ancoras-destorcedores-cabo-de-polipropileno-esticadores-correntao-maritimo>. Acesso on: October 31, 2016.

MONGABAY. Online. Available : <http://pt.mongabay. com/news/2012/pt0223-conservation drone.html>. Accessed on July 7, 2014.

ENVIRONMENTAL WORLD. APP and Legal Reserve. 2010. Online. Available at: < http ://mundoambiental2010.blogspot.com. br/>. Accessed on October 15, 2016.

OCORRENTAO .Online. Available at : <http://www.ocorrentao.com.br/?lightbox=dataItem-inf3nqly2> Accessed on October 14, 2016.

NEW ROLAND. D130. Catalog. 10p .2009. Online. Available at: http : //www.batioli.com.br/ad/arquivo/catalo go/D 13 0 Crawler tractor. pdf. Accessed on October 23, 2016.

REVEGETATION EQUIPMENT CATALOG. Controlling Plants Mechanically. Online. Available at: < http://reveg-catalog.tamu.edu/04- Mechanical.htm>. Accessed on October 28, 2016.

ROME. Land clering and maintenance equipment. Online. Available at: < http://www.romeplow.com/KG%20and%20VEE%20Blades.htm>. Accessed on October 23, 2016.

SAAD, O. - Machines and techniques for initial soil preparation. Sao Paulo, Nobel, 1977.

SANTOS FILHO, A. G.; SANTOS, GUARNETI, J. E. G. S. Apostila de màquinas agricolas. Bauru, 88 p. 2001.

SOBIOLOGY. The two major groups of angiosperms. Online. Available at: < http://www.sobiologia.com.br/conteudos/Reinos4/angiospermas2.php>.

Accessed on: October 15, 2016.

TESTA, A. Mechanization of deforestation: the new frontiers. Sao Paulo, Ed.Agrônomica Ceres, 314 p., 1983

TERRAMOV. Online. Available at: < http://www.terramov.com.br/servicos.php?imgs=8>. Accessed on: October 31, 2016.

THORCO. Online. Available at: < http://www.thorco.com.br/produtos.php?cat=3&produto=81 >. Accessed on: 31 Jan. 2017.

VENANCIO, D. L. et al. Mapping *auracaria angustifolia* using GPS receivers. Cerne, Lavra, v. 16, n.3, p. 391-398, 2010.

WORD METERS. Online. Available at: <http : //www.worldometers .info/br/>. Accessed on July 7, 2014.

WWF. Riparian forest conservation. Online. Available at: http://www.wwf.org.br/natureza brasileira/reducao de impactos2/agricultura/a gr acoes resultados/agr solucoes mata ciliar/ Accessed on: Oct. 15, 2016.

XLHJ GROUPLIMITED . Online. Available at: <

http://en.beijingxlhj .com/SHANTUI-SD 16>. Accessed on: October 16, 2016.

Printed by Books on Demand GmbH, Norderstedt / Germany